Meditations on the Rainbow

poetry by

Sapphire

Copyright ©1987 by Sapphire

All rights reserved. No part of this book may be reproduced or transmitted in any form by any means, electronic or mechanical including photocopy and recording, or by any information storage or retrieval system, without permission in writing from Crystal Bananas Press, P.O. Box 975, Manhattanville Station, New York, NY 10027.

ISBN 0-931885-00-0

Cover, design & production: Sorah Broder

Editorial consultant & proofreading: Carletta J. Walker

Back cover photograph: Annette Pelaez

 Produced at The Print Center., Inc., 225 Varick St., New York, NY 10014, a non-profit facility for literary and arts-related publications. (212) 206-8465

I would like to acknowledge and thank all the women who donated money for the publication of this book.

I would also like to thank Sorah Broder and Carletta J. Walker for their loving and invaluable assistance in bringing this book into being.

*this book is dedicated to
all the homosexuals, lesbians, queers,
faggots, dykes, fairies, zami queens,
jaspers, wimmin lovers and bulldaggers
of the rainbow*

Contents

yellow

brain
cells
activated,
messages
running,
flying
lemon drops
over synapse,
yellow is the thot force.
We need
to think
about
nuclear waste
disposal
& since
we LUV
Africa
so much
we
need to
figure out
how to stop
the crackers
from dumping
lethal
radioactive
wastes
there,
first
we gotta
stop em
from riding
thru Harlem
wit the shit.
We need
to think

about
aluminum
poisoning
from the
pots
we cook
in
& how to stop
ourselves from
mad consumerism,
how to give
ourselves
enuff
orgasms
cornbread
essays
poems
warm baths
& inner peace
so we stop
feeling
the only way
to feel
some power
is to
buy something.
Think now,
lets think
how
we can
free
our minds
from this
homophobia
fear of difference
passive acceptance of

men
as
rulers
leaders
providers
destroyers,
lets
come up
with a plan
on
how to
stop
the building
of more
nuclear plants
& how to
bring about
the shut
down
of
existing
ones.
Think now
how to
reintegrate
ourselves
into our
Black woman
selves
& be
the bond
warp & weft
knit ourselves
together
in a
new

thot out
unity
so we
dont
starve
to death
or end up
sleeping
on the
subway.
Lets find out
what we
need
& band
together &
dance & sing
for each other,
lets learn
how to fight
how to document
our existence
so no more
children
will come
behind
us
thinking
we didnt
exist,
think
zami queens
lesbians
wimmin lovers
jaspers
dykes
homosexuals

bulldaggers
didnt
exist,
think
we was
a wite disease
instead of a purple
corundum of black &
lavender power
that has filled
our people
with
ideas
music
literature
babies
brains &
beauty.
Let us
not hide
behind
euphemisms
but embrace
the despised
& analyze
& find
& share
& BE
who we
are.
Think
on
it.

red

primal
power
squat
scream
blare
bellow
RAGE
shake
dance
deny
fulfill
PASSION
strong
sinews
dripping
tits
squat
grunt
push
push
blood
new
life
blood
power
primal
power
birth
birth
ahh!
scream
new
life

black

black
magic
dread
magic
rhythm
ri THUM
rhythm
the
force,
the
primal
force,
the
encompassing
of all,
the
beginning
& the
end,
black
soft
smooth
succulent
suck
like
sweet
black
grapes
that
grand
fat
Black
woman
in
her
fuchsia

purple
lipstick,
Essence
hairdo
& turquoise
unitard,
so
sweet
to
suck
a
glory
untold
story
of
a
fat
sumptuous
Black
woman
of
large
hips
& able
breast,
flying
tongue,
earthbound
rain
forest
in her
mind,
a
universe
in her
bowels,

miracles
in her
ovaries,
flying
fat
full
thru the
Rockies
top of
Kilimanjaro
a
molten
force
coursing
thru the
earth
a
burning
greedy
core
spewing
hot
lava
burning
raw
peeling
the
skins
off
She
created
the
pink
race.
Mother
of

us
all
sits
in the
primordial
squat
kills
kills
gives
& takes,
eyes
glowing
red
like
a
traffic
lite,
mucus
yellow
secretions
odorous
& wonderful
& foul
slide from
her
womb
her
whole,
black
hole
crater
in
space
original
universe
& god

is
the
clit
a
burning
red
projectile
amidst
the
damp
warm
secretions
of
creation.
Black
mother
lover
sister
I
fall
onto
you
with
arms
embracing,
my
mouth
open
wide
to
suck
give
lick
receive
love
willing

to
be
all
capable
of
all
am
all
I
you
join
fire
lite
wet
funky
conjunction
never
before
like
now.
Oww
witch.
Fat
Black
witch,
I
am
not
afraid.
I
love
you.

lavender

counter
culture
gay
girls,
the
straight
women
on the
bus
move
away
from
us.
We
stand
in
lavender
the
force
society
demand
we
hide,
lavender
rebellion
homosexual
persuasion.
HOMOSEXUAL,
there it is,
that
divine
feared
word,
love
your
own,

same
sex,
penis
to
anus,
tongue
to
vagina,
the
boys
&
the
boys,
the
girls
&
the
girls,
fields
of
blue
flowers
&
fields
of
pink
flowers
turning
to
LAVENDER
stars
triangles
crosses,
we have
been
gassed,

the Bible
has
damned
us,
colored people
say
we are
of
European
origin,
Europeans
claim
we are
of
the
devil.
We exist
we exist,
lavender scents
grown
strong
in
the
midst of
holocaust &
nuclear
disaster,
an
ever
present
beagle nose
witch
hunt
goes
on
to

stamp
us
out
but
we
grow
strong,
clone,
mirror
the
oppressor
in
leather &
jeans,
cowboys &
big bad
butches
fill
the
streets
with
blood,
the
Lone
Ranger
is
dead &
Tonto
is
no
fool.
We
are
on our
own now,
color & class

divide
us,
we
seep
miasma
mauve
clouds
away from
each
other,
separated
from
our
common
bond
of
queerness,
deviant
difference
& sweet divinity,
separatéd
by
whips
swastikas
& the
gay wite
gentrification
that sent
poor Black
families
in San Francisco
out
of
their
homes of
generations

in the
Filmore district
to the
outskirts
of
San Francisco &
Daly City
as clones
marched
thru
the Filmore & Haight
to the tune of
            this land is your land
            this land is my land
            from california to the
            new york island
wite
wite
imperialism,
take
what
you
want.
A
young
Black man
shoots
thru
the
window
of
a
Christopher Street
bar
screaming
as

he kills
the
boy
who
last
sucked
his
dick
>   *i'm not gay*
>   *i'm not gay*
>   *i'll never say i'm gay*
sick
sick
internalized
homophobia.
Petit bourgeois
opportunists
use
the community
but dont
let
the community
use them,
queen hip
grand dreads
are only
gay
when
its
convenient
& when
it
aint
convenient
they
can be

heard
talkin
bout
they are
human beings
& dont
want
to
be
labled
or
confined
to
a
box.
But *still*
in spite of
racism,
classism,
internalized homophobia &
petit bourgeois opportunism
a
lavender
mist
is
rising,
filling
the
cavities
of the
brain,
raining
rivers
of
vaginal
fluids,

waving
wild
clits
in the
air
&
one
breasted
warriors
ride
silently
calling out
in the nite
SAVE OUR SISTERS
SAVE OUR SISTERS

wite

the wite
glows
as a
message
a reminder that
the first shall be last
things change
& the
snow queen
didn't hafta
be so
mean
there was
enuff for
everybody
& big monolith
ass women
from way back
B.C.
lost power
to men
& thats when
a lot of this
shit began,
the wite
glows
as a
belief
even they
can
change.

green

forest
green,
grass
green,
healing live
green,
clean green.
Green is for the land
they stole/are stealing.
Palestine,
South Africa,
the Americas—
North, South & Central
from the Indians,
Palestine
from the Palestinians,
South Philly & Center City
in Philadelphia
from the Blacks;
Park Slope, Harlem,
the Lower East Side
bits & pieces
a brownstone here
a building there,
middle class wites
on the flight back
from Long Island
Connecticut &
Queens
risking life
in a
'changing
neighborhood',
so
courageous
open-minded

accepting
eating up
the
house
building
for
50
500
5000 dollars
while
the residents,
lifelong
tenants,
who cannot
afford
or are not
given
the opportunity
to buy
are piled up
in projects
way out in
west hell or
Far Rockaway.
And the crackers
strip the
bannisters
doors
woodwork
& floors,
'LOOK what we
found under ALL
that linoleum!'
they cry
in middle class
pride

as they
show off
their new
freshly stained
& polyurethaned
floors,
catching cabs
to & from
their doorsteps,
afraid of the
'unreasonable
resentment of some
of the occupants
of the neighborhood,
after all
it WAS
going down
the drain
till WE
came...'
The wite middle class
took the Filmore &
Haight Ashbury districts
in San Francisco, California
they happened
to be gay,
the wite middle class
took Park Slope
in Brooklyn, New York
they happened
to be heterosexual,
what I'm saying is:
its a class issue
New York
Puerto Rico
Hawaii

Alaska
land
land
wite people
eat land
oil
other people
property owners
owners of the earth
Azores
Maori Islands.
The Indians say
no one OWNS the land,
when they fought
like the
Palestinians
they were
called savages
& terrorists too,
'Terrorists,
underdeveloped
cavemen
should be *glad*
we've come
to take
the land,
we suffered
we Pilgrims & Puritans
we suffered
religious persecution
in the
old country,
filthy savages
werent doing
anything
with the land

anyway'
(But LIVING on it!)

> *this land is your land*
> *this land is my land*
> *from california*
> *to the new york island*

This land
was the
Indian's land!
Stolen with
beads &
blood
gun powder
connivery
thievery
& religious zeal.
Pieter W. Botha,
prime minister of
South Africa,
and all the
crackers in Europe
& the United States
called the Black
guerillas fighting
for their land
in Africa
savages,
terrorists.
Listen little Black
brown children
it is a
good thing
to be
called a terrorist
in that
context,

it is bad
to be like
Hitler, Reagan
or Begin,
it is bad
to be a
Western European
wite supremist
taking land
from people
of color
who have
lived
on it
for centuries
TWO WRONGS DONT MAKE A RIGHT
cause the cracker
kill the jew
dont mean
he should come take
the land
from you
LONG LIVE THE P.L.O.
restitution for the
Native Americans,
give the land
back
to the Arapaho
the Souix
Crow
Apache
Hopi
Navaho
Miami
Seminoles
Creeks.

People of color
stand for all other
peoples of color,
what happens to Blacks
will happen to Native Americans
Palestinians, Asians (all the absorption
of wite western culture by the Japanese
didnt stop the United States from
dropping the bomb on Hiroshima
& Nagasaki).
Nuclear war
will kill.
It will
kill
us.
Be responsive
to ourselves,
fight for our
planet's
life,
our earth
the grass
the sweet trees
flowers
ocean
children
have all
been
hurt,
we have
been
reduced
as
human beings,
our
vision blurred.

We can control
our destiny
this hell
on earth
does not
have to
be.
Green life
green is for the land
shoots of rice
bamboo
corn
cassava
spring time
green
live
life
I want to live!
I want the world to live!
see green
see green
sea green

blue

Big Mama Thornton
is dead,
aggressive
blues shouter
wild woman
attack heart broke
burnt out liver
she could blow
that mouth harp,
she loved the blues
& women.
I see Leza
in that woman's
haunting spirit
walk
touching
everything
holding
the real goods
close
real good woman
reeling to a
deep down
funky butt
sound,
the blues is a
lifelong tangent
of serious love
twisted
transformed
broken
glued together
melted
bought
stolen
sold

packaged
boxed
marketed
damned
sliced
dismembered
mutilated
raped
asshole
split
stomach
snatched
outta
throat
eyes
eliminated
tubes
tied
hair
a tragedy
a tragedy
what done happpened
to our hair

the sky turns
to midnite dusk
all shades of blue
all blues
all blue
how long
howl on
HOWL ON
The Wolf
streaking down
ice streets
in Illinois

South Side of
Chicago

I swigged that
Courvoisier
down my throat
& wrapped
my wild young
thighs
around your
back
like a pretzel
my lips
around your
dick
like a suction
tube
& I pulled
begging release
from this
ball & chain

somethin somethin!

you hear me
somethin.
I was alrite
for awhile

*sittin by my window
lookin out lookin out
at the rain
& somethin came along
& grabbed a hole of me
& it felt jus like
a ball & chain***

Ahh!
i went after
my own self
wit razor blades
& all my teeth
fell out when
i opened that letter
said another lover
i live now with
another lover
i hope you
are happy
now,
i am.
ahh!
something tole me it was over
no one sings that like Etta James
when i saw you and that girl walking
something deep down in my soul said cry girl
i wanna snatch you out
my heart, my vision
i rather
go head girl
i rather go blind than see
*you walk away from me**
why did you leave
something deep down in my soul
all the poems
all the dances
all the everything
cannot make my soul sing
the way you did sweet girl
two weeks has ate up two years
of my life now
sweet pearls raining from the sky
turn to moons between my legs

for you to swallow
now listen
i cannot deal with
never seeing you again
but i hafta
other people bounce back
from these things
why cant i bounce back
blue black strong
thru delta mud, juke joints, cotton fields
places i ain never been,
i only been in a cotton field once
& that was when i was 19 & ran away
from Synanon with this spozed to be bad
bitch from Detroit who turned out to not wanna
steal *or* hoe & suggested we *work* for whatever cash
we needed & we ended up in Fresno chopping cotton
under the hot sun.
i never did that again,
the old folks say you dont know *what*
you comin to 'for you die
but *i* know i aint goin back to that!

i wanna ride a wite horse
thru the midnite
dig spurs in her side
hear her cry
till i am thrown
off into the midnite again

> *you got me runnin*
> *you got me hidin*
> *any way you want it**

i will deliver
me from this
loneliness
muddy

muddy water
muddy waters
Muddy Waters
deep delta sound

> *i dont want you to*
> *bake my bread*
> *i dont want you to*
> *make my bed*
> *i dont want you*
> *sad & blue*
> *i just wanna make*
> *luv to you   luv to you*
> *luv to you. . . ***

Chicago sound, Black people, blues
The Wolf, Muddy, Little Walter
unknown, unheard beyond the Pepper Lounge
& Teresa's on the South Side of Chicago
soon their imitators would command
$10,000 per performance &
Big Mama Thornton would live
once more in wite face
this time a woman, Janis Joplin.
Etta James, L.A. woman, would be
rehearsing in them clubs during the day
getting ready for them nite gigs
& day after day
week after week
month after month
would be this
same lil ol' wite girl
sittin at a table
in the back,
no matter which club
she be in
same lil ol' wite bitch

sittin listenin.
Finally one day
Etta say, ''Who the fuck is that?''
One of the side men tell her
''Oh that's Janis Joplin.''
I was at the Palladium
for James Brown's
25th Anniversary Show
& two boys was
in back of me
one Black, the other wite
they was together
a couple
& we was all
sittin listenin
to Etta James
who was opening
up the show
& the wite boy leaned
over to the Black one
& said, ''She sounds just like Janis Joplin!''
America
is a
mutherfucker
the
cultural
imperialism
has been
as
vast
deathly
nasty
lo down
& ugly
as any other
kind of imperialism,

the practice
of conquering
a people,
taking their
raw goods &
selling them
back to
them
has happened
in the
music industry
too
& lemme tell ya
the wite girls
do it
as good as
the men!
Yeah I'm mad.

when we moved from Texas to Philadelphia
the kids usta make fun of my southern accent
hurt me so bad
made me ashamed of where i came from
but i can tell you what was tole me,
the blues aint nuthin to be ashamed of
they are the roots of all American music,
get that,
all
any & all
American music
has its roots
in Black people,
we are the spirit & soul
of this nation,
the only thing
saving it from

pure & total
depravity,
the only thing America
can be proud of
is its music,
we cannot take
pride in
Hiroshima
Vietnam
Campbells Soup
Andy Warhol
The Trail of Tears
Wounded Knee
Three Mile Island
The Middle Passage
Liza Minelli
Charles Manson
Doris Day or
Cary Grant

one of the first things the Beatles said
when they got here was that they wanted
to see Muddy Waters & Bo Diddley.
a reporter asked, "where's that?"
they laffed and said, "dont you know
who your own famous people are here?"

corn bread, blackeyed peas, army rations,
beef jerky, red #2, M&M's
i was from working class colored folks
who wanted to do better
i did not eat a mango
till i was 26
now *that* is cultural deprivation
a thousand tears
spread out over

my sweetness lost
irretrievably stuffed
in a trash can
Karen, my best friend,
found dead
in a
trash can,
tears tore into
my flesh
like a
rubber hose
beat
beat
the hose
searing
my flesh,
i did not know
that was not
par for the course
till i was
grown,
getting beat
like
a dog
ass bitch
snag a tooth hoe
nasty ass
bulldagger
douch bag mouf
wench
slut
sleazy ass
tramp
whore
hole
hoe

nuthin
never was
never will be
nuthin
my foot lives
up your
ass
inferior being
woman
got to bleed
every month
& have
babies
babies
blue babies
stillborn babies
disabled babies
cesarean babies
kwashiorkor,
disease of the first child
when the second one
is born,
starvation,
after the breast
there is precious
little else
in the way of protein,
marasmus
harlem, little africa
they die
like flies
babies,
soda pop &
reefer
do not create
healthy

babies
babies
precious
little
balls of
life
die so easily
cigarette
burns
will kill
them,
do you know what it feels like
to be little, lying in your crib
wet, scratchy, a pang in your
empty stomach, crying.
crying cause you have learned
crying brings results
warm dryness between the legs
& the warm plastic
nipple full of sweet milk.
but this time it is the nasty acrid smell
that is sometimes on her lips
coming close to you hot
HOT into your skin
& you scream & scream
& the more you scream
the more she burns,
you are little, you are tiny
nothing you could have done warrants this
but you dont know this
& *if* you live
you grow up with that memory
buried in your bowels
a quiet recording telling you
you are nothing
nobody

Baby baby
I'm for *real*
just let me put the tip in
no nooo *NO*!
back seat rapes
this date
is nothin like what
the teen scene reporter in
Seventeen magazine
said your first date would be like
baby baby
what you see,
misogynist
is what
you get

encompass my ear
in your mouth
& make me holler
all down Central Avenue
for you
& you know
i will
cause i'm just
that kind of woman
love is
an indigo mask
i put on
before i
ride
& become
numerous people
including
myself
& dont get
mixed up

with me
or the blues
less u ready
able
to get
down
all
the
way
down
flying
thru the
bowels of
12 string
guitars
wash boards
& wang dang doodle
unh unh UH
chile chile *chile*
little girl
little girl
u dont know
what chu do
to me.

page 59
* as sung by Howlin Wolf
** as sung & written by Willie Mae "Big Mama" Thornton

page 60
* as sung by Etta James

page 61
* as sung by Jimmy Reed

page 62
* as sung by Muddy Waters, written by Willie Dixon

Page 64
in 1983 at the Long Beach Blues Festival Willie Dixon told the audience, "...the blues is the roots of *all* American music!"